Tucholsky Wagner Zola Scott Sydow Freud Schlegel
Turgenev Wallace Fonatne
Twain Walther von der Vogelweide Fouqué Friedrich II. von Preußen
Weber Freiligrath Frey
Fechner Fichte Weiße Rose von Fallersleben Kant Ernst Richthofen Frommel
Hölderlin
Engels Fielding Eichendorff Tacitus Dumas
Fehrs Faber Flaubert
Maximilian I. von Habsburg Fock Eliasberg Zweig Ebner Eschenbach
Feuerbach Ewald Eliot Vergil
Goethe London
Elisabeth von Österreich
Mendelssohn Balzac Shakespeare Dostojewski Ganghofer
Lichtenberg Rathenau Doyle Gjellerup
Trackl Stevenson Hambruch
Mommsen Tolstoi Lenz Hanrieder Droste-Hülshoff
Thoma von Arnim Hägele Hauff Humboldt
Dach Verne
Reuter Rousseau Hagen Hauptmann Gautier
Karrillon Garschin Defoe Hebbel Baudelaire
Damaschke Descartes
Hegel Kussmaul Herder
Wolfram von Eschenbach Dickens Schopenhauer Rilke George
Bronner Darwin Melville Grimm Jerome Bebel Proust
Campe Horváth Aristoteles Voltaire Federer
Bismarck Vigny Barlach Herodot
Gengenbach Heine
Storm Casanova Tersteegen Gilm Grillparzer Georgy
Chamberlain Lessing Langbein Gryphius
Brentano Claudius Schiller Lafontaine
Strachwitz Kralik Iffland Sokrates
Katharina II. von Rußland Bellamy Schilling
Gerstäcker Raabe Gibbon Tschechow
Löns Hesse Hoffmann Gogol Wilde Gleim Vulpius
Luther Heym Hofmannsthal Klee Hölty Morgenstern Goedicke
Roth Heyse Klopstock Puschkin Homer Kleist
Luxemburg La Roche Horaz Mörike Musil
Machiavelli Kierkegaard Kraft Kraus
Navarra Aurel Musset Lamprecht Kind Moltke
Nestroy Marie de France Kirchhoff Hugo
Nietzsche Nansen Laotse Ipsen Liebknecht
Marx Lassalle Gorki Klett Leibniz Ringelnatz
von Ossietzky May vom Stein Lawrence Irving
Petalozzi Platon Knigge
Sachs Pückler Michelangelo Kock Kafka
Poe Liebermann Korolenko
de Sade Praetorius Mistral Zetkin

The publishing house tredition has created the series **TREDITION CLASSICS**. It contains classical literature works from over two thousand years. Most of these titles have been out of print and off the bookstore shelves for decades.

The book series is intended to preserve the cultural legacy and to promote the timeless works of classical literature. As a reader of a **TREDITION CLASSICS** book, the reader supports the mission to save many of the amazing works of world literature from oblivion.

The symbol of **TREDITION CLASSICS** is Johannes Gutenberg (1400 – 1468), the inventor of movable type printing.

With the series, tredition intends to make thousands of international literature classics available in printed format again – worldwide.

All books are available at book retailers worldwide in paperback and in hardcover. For more information please visit: www.tredition.com

tredition was established in 2006 by Sandra Latusseck and Soenke Schulz. Based in Hamburg, Germany, tredition offers publishing solutions to authors and publishing houses, combined with worldwide distribution of printed and digital book content. tredition is uniquely positioned to enable authors and publishing houses to create books on their own terms and without conventional manufacturing risks.

For more information please visit: www.tredition.com

A Queens Delight The Art of Preserving, Conserving and Candying. As also, A right Knowledge of making Perfumes, and Distilling the most Excellent Waters.

W. M.

Imprint

This book is part of the TREDITION CLASSICS series.

Author: W. M.
Cover design: toepferschumann, Berlin (Germany)

Publisher: tredition GmbH, Hamburg (Germany)
ISBN: 978-3-8491-6652-6

www.tredition.com
www.tredition.de

Copyright:
The content of this book is sourced from the public domain.

The intention of the TREDITION CLASSICS series is to make world literature in the public domain available in printed format. Literary enthusiasts and organizations worldwide have scanned and digitally edited the original texts. tredition has subsequently formatted and redesigned the content into a modern reading layout. Therefore, we cannot guarantee the exact reproduction of the original format of a particular historic edition. Please also note that no modifications have been made to the spelling, therefore it may differ from the orthography used today.

Table of Contents

A QUEENS DELIGHT OF Conserves, and Preserves, Candying and Distilling

To preserve white Pear Plums, or green.

To preserve Grapes

To preserve Quinces white.

To preserve Respass.

To preserve Pippins.

To preserve fruits green.

To preserve Oranges and Lemons the best way.

An approved Conserve for a Cough or Consumption of the Lungs.

To make conserve of Any of these Fruits.

To dry any Fruits after they are preserved, to or Candy them.

To preserve Artichokes young, green Walnuts and Lemons, and the

To preserve Quinces white or red.

To preserve Grapes.

To preserve Pippins, Apricoks, Pear-Plums and Peaches when they are

To preserve Pippins, Apricocks, Pear-Plums, or Peaches green.

To dry Pippins, or Pears without Sugar.

To make Syrup of Clove-gilly flowers.

To make Syrup of Hysop for Colds.

To make Orange Water.

To dry Cherries.

To make juyce of Liquorish.

A Perfume for Cloths, Gloves.

To make Almond Bisket.

To dry Apricocks.

To make Quinces for Pies.

The best way to break sweet Powder.

To make excellent Perfumes.

To make Conserve of Roses boiled.

To make Conserves of Roses unboiled.

To make a very good Pomatum.

To make Raisin Wine.

To make Rasberry Wine.

The best way to preserve Cherries.

A Tincture of Ambergreece.

To make Usquebath the best way.
To preserve Cherries with a quarter of their weights in Sugar.
To make Gelly of Pippins.
To make Apricock Cakes.
To preserve Barberries the best way.
To make Lozenges of Red Roses.
To make Chips of Quinces.
To make Sugar of Wormwood, Mint, Anniseed, or any other of that kinde.
To make Syrup of Lemons or Citrons.
To make Jambals of Apricocks or Quinces.
To make Cherry-water.
To make Orange Cakes.
To preserve Oranges the French way.
To preserve green Plums.
To dry Plums.
To preserve Cherries the best way, bigger than they grow naturally,
To preserve Damsins, red Plums or black.
To dry Pippins or Pears.
To dry Pippins or Pears another way.
To dry Apricocks tender.
To dry Plums.
To dry Apricocks.
Conserves of Violets the Italian manner.
Conserves of red Roses the Italian manner.
Conserve of Borage Flowers after the Italian manner.
Conserve of Rosemary flowers after the Italian manner.
Conserve of Betony after the Italian way.
Conserve of Sage.
Conserve of flowers of Lavender.
Conserve of Marjoram.
Conserve of Peony after the Italian way.
Touching Candies, as followeth.
To Candy Rosemary-flowers in the Sun.
To Make Sugar of Roses.
To Candy Pippins, Pears, Apricocks or Plums.
To Candy or clear Rockcandy flowers.
To Candy Spanish Flowers.
To Candy Grapes, Cherries or Barberries.
To Candy Suckets of Oranges, Lemons, Citrons, and Angelica.

To Candy the Orange Roots.
Candy Orange Peels after the Italian way.
To Candy Citrons after the Spanish way.
Candied Cherries, the Italian way.
Chicory Roots candied the Italian way.
Touching Marmalets, and Quiddony, as followeth.
To make Marmalet of Damsins.
To make white Marmalet of Quinces.
To make Marmalet of any tender Plum.
To make Orange Marmalet.
To make Quiddony of Pippins of Ruby or any Amber colour.
To make Quiddony of all kind of Plums.
To make Marmalet of Oranges, or Orange Cakes, &c.
Touching Pastrey and Pasties.
To_make_Sugar_Cakes.
To make clear Cakes of Plums.
To make Paste of Oranges and Lemons.
To make Rasberry Cakes.
To make Paste of Genoa Citrons.
To make a French Tart.
To make Cakes of Pear Plums.
To make Cakes, viz.
To make a Cake the way of the Royal Princess, the Lady Elizabeth,
To make Paste of Apricocks.
To make Paste of Pippins like leaves, and some like Plums, with their
To make Paste of Elecampane roots, an excellent remedy for the Cough of
To make Paste of flowers of the colour of Marble, tasting of natural
To make Paste of Rasberries or English Currans.
To make Naples Bisket.
To make Italian Biskets.
To make Prince Biskets
To make Marchpane to Ice and Gild, and garnish it according to Art.
Lozenges
To make Walnuts artificial.
To make Collops like Bacon of Marchpane.
To make artificial Fruits.
Touching Preserves and Pomanders.

To make an excellent perfume to burn between two Rose leaves.
To make Pomander.
To make an Ipswich Water.
To make a sweet Smell.
Touching Wine.
To make Hypocras.
The Lady **Thornburghs** *Syrup of Elders.*
To make gelly of Raspis the best way.
To dry Fox Skins.
Choice Secrets made known.
To make true Magistery of Pearl.
How to make Hair grow.
To write Letters of Secret, that they cannot be read without the
How to keep Wine from Sowring.
To take out Spots of Grease or Oyl.
To make hair grow black, though any colour.
King **Edwards** *perfume.*
Queen **Elizabeths** *Perfume.*
Mr. **Ferene** *of the* **New Exchange,** *Perfumer to the Queen, his rare*
To make the said Powder into Paste.
The Receipt of the Lady **Kents** *powder, presented by her Ladyship to*
A Cordial Water of Sir **Walter Raleigh.**
The Lady **Malets** *Cordial Water.*
A Sovereign Water of Dr. **Stephens,** *which he long times used,*
A Plague Water to be taken one spoonful every four hours with one sweat
Poppy water.
A Water for a Consumption, or for a Brain that is weak.
Another of the same.
A good Stomach Water.
A Bag of purging Ale.
The Ale of Health and Strength, by Viscount St. **Albans.**
A Water excellent good against the Plague.
A Cordial Cherry-water.
The Lord **Spencers** *Cherry-water.*
The Herbs to be distilled for Usquebath.
Dr. **Kings** *way to make Mead.*
To make Syrup of Rasberries.
To make Lemon Water.

To make Gilly-flower Wine.
The Lady **Spotswood** *Stomach Water.*
Water of Time for the Passion of the Heart.
A Receipt to make damnable Hum.
An admirable Water for sore Eyes.
A Snail Water for weak Children, and old People.
Clary Water for the Back, Stomach, &c.
Dr. **Montfords** *Cordial Water.*
Aqua Mirabilis, Sir **Kenelm Digby's** *way.*
A Water for fainting of the Heart.
A Surfeit Water.
Dr. **Butlers** *Cordial Water against Melancholly, &c. most approved.*
The admirable and most famous Snail Water.
A singular Mint water.
Distillings.
A most Excellent **Aqua Coelestis** *taught by Mr.* **Philips** *Apothecary.*
Hypocras taught by Dr. **Twine** *for Wind in the Stomach.*
Marigold flowers distilled, good for the pain of the Head.
A Water good for Sun burning.
The Lady **Giffords** *cordial Water.*
A water for one pensive and very sick, to comfort the Heart very
To perfume Water.

A QUEENS DELIGHT OF Conserves, and Preserves, Candying and Distilling Waters.

To preserve white Pear Plums, or green.

Take the Plums, and cut the stalk off, and wipe them then take the just weight of them in Sugar, then put them in a skillet of water, and let them stand in and scald, being close covered till they be tender, they must not seeth, when they be soft lay them in a Dish, and cover them with a cloth, and stew some of the the Sugar in the glass bottom, and put in the Plums, strewing the sugar over till all be in, then let them stand all night, the next day put them in a pan, and let them boil a pace, keeping them clean scummed, & when your Plums look clear, your syrup will gelly, and they are enough. If your Plums be ripe, peel off the skins before you put them in the glass; they will be the better and clearer a great deal to dry, if you will take the Plums white; if green, do them with the rinds on.

To preserve Grapes

Take Grapes when they be almost through ripe, and cut the stalks off, and stone them in the side, and as fast as you can stone them strew Sugar on them; you must take to every pound of Grapes three quarters of a pound of Sugar, then take some of the sower Grapes; and wring the juyce of them, and put to every pound of Grapes two spoonfuls of juyce, then set them on the fire, and still lift up the pan and shake it round, for fear of burning to, then set them on again, & when the Sugar is melted, boil them as fast as you can possible, and when they look very clear, and the syrup is somewhat thick, they are enough.

To preserve Quinces white.

Take a pair and coar them, and to every pound of your equal weights in Sugar and Quince, take a wine pint of water; put them together, and boil them as fast as you can uncovered; and this way you may also preserve Pippins white as you do Quinces.

To preserve Respass.

Take a pound of Respass, a pound of fine Sugar, a quarter of a pint of the juyce of Respass, strew the Sugar under and above the Respass, sprinkle the juyce all on them, set them on a clear fire, let them boil as soft as is possible, till the syrup will gelly, then take them off, let them stand till they be cold, then put them in a glass. After this manner is the best way.

To preserve Pippins.

Take fair Pippins, and boil them in fair water till they be somewhat tender, then take them out, and peel off the skins and put them into a fair earthen pot, and cover them till they be cold, then make the syrup with fair water and Sugar, seeth it, and scum it very clean, then being almost cold, put in your Pippins, so boil them softly together, put in as much rind of Oranges as you think will tast them, if you have no Oranges take whole Cinamon and Cloves, so boil them high enough to keep them all the year.

To preserve fruits green.

Take Pippins, Apricocks, Pear-Plums, or Peaches when they be green, scald them in hot water, and peel them or scrape them, put them into another water not so hot as the first, then boil them very tender, take the weight of them in Sugar, put to it as much water as will make a syrup to cover them; then boil them something leisurely, and take them up, then boil the syrup till it be somewhat thick, that it will batten on a dish side, and when they are cold, put them together.

To preserve Oranges and Lemons the best way.

Take and boil them as for paste, then take as much sugar as they weigh, and put to it as much water as will cover them by making a syrrup, then boil them very leisurely till they be clear, then take them up and boil the syrup till it batten on the dish side, and when they are cold put them up, &c.

An approved Conserve for a Cough or Consumption of the Lungs.

Take a pound of Elecampane Roots, draw out the pith, and boil them in two waters till they be soft, when it is cold put to it the like quantity of the pap of roasted Pippins, and three times their weight of brown sugar-candy beaten to powder, stamp these in a Mortar to a Conserve, whereof take every morning fasting as much as a Walnut for a week or fortnight together, and afterwards but three times a week. *Approved.*

To make conserve of Any of these Fruits.

When you have boiled your paste as followeth ready to fashion on the Pie-plate, put it up into Gallipots, and never dry it, and this is all the difference between Conserves. And so you may make Conserves of any Fruits, this is for all hard Fruits, as Quinces, Pippins, Oranges and Lemons.

To dry any Fruits after they are preserved, to or Candy them.

Take Pippins, Pears or Plums, and wash them out in warm water from the syrup they are preserved in, strew them over with searsed Sugar, as you would do flower upon fish to fry them; set them in a broad earthen Pan, that they may lie one by one; then set them in a warm Oven or Stove to dry. If you will candy them withall, you must strew on Sugar three or four times in the drying.

To preserve Artichokes young, green Walnuts and Lemons, and the Elecampane-Roots, or any bitter thing.

Take any of these, and boil them tender, and shift them in their boyling six or seven times to take away their bitterness out of one hot water into another, then put a quart of Salt unto them, then take them up and dry them with a fair cloth, then put them into as much clarified Sugar as will cover them, then let them boil a walm or two, and so let them stand soaking in the Sugar till the next morning, then take them up and boil the Sugar a little higher by it self, and when they are cold put them up.

Let your green Walnuts be prickt full of holes with a great pin, and let them not be long in one water, for that will make them look black; being boiled tender, stick two or three Cloves in each of them.

Set your Elecampane-Roots, being clean scraped, and shifted in their boilings a dozen times, then dry them in a fair cloth, and so boil them as is above written, take half so much more than it doth weigh, because it is bitter, &c.

To preserve Quinces white or red.

Take the Quinces, and coar them, and pare them, those that you will have white, put them into a pail of water two or three hours, then take as much Sugar as they weigh, put to it as much water as will make a Syrup to cover them, then boil your Syrup a little while, then put your Quinces in, and boil them as fast as you can, till they be tender and clear, then take them up, and boil the Syrup a little higher by it self, and being cold put them up. And if you will have them red, put them raw into Sugar, and boil them leisurely close covered till they be red and put them not into cold water.

To preserve Grapes.

Take the Clusters, and stone them as you do Barberries, then take a little more Sugar than they weigh, put to it as much Apple water as will make a Syrup to cover them, then boil them as you do Cherries as fast as you can, till the Syrup be thick and being cold pot it, thus may you preserve Barberries or English Currans, or any kind of Berries.

To preserve Pippins, Apricoks, Pear-Plums and Peaches when they are ripe.

Take Pippins and pare them, bore a hole through them, & put them into a Pail of water, then take as much Sugar as they do weigh, and put to it as much water as will make a Syrup to cover them, and boil them as fast as you can, so that you keep them from breaking, until they be tender, that you may prick a Rush through them: let them be a soaking till they be almost cold, then put them up.

Your Apricoks and Peaches must be stoned & pared, but the Pear-Plums must not be stoned nor pared. Then take a little more Sugar than they weigh, then take as much Apple water and Sugar as will make a Syrup for them, then boil them as you do your Pippins, and Pot them as you do the Pippins likewise, &c.

To preserve Pippins, Apricocks, Pear-Plums, or Peaches green.

Take your Pippins green and quoddle them in fair water, but let the water boil first before you put them in, & you must shift them in two hot waters before they will be tender, then pull off the skin from them, and so case them in so much clarified Sugar as will cover them, and so boil them as fast as you can, keeping them from breaking, then take them up, and boil the syrup until it be as thick as for Quiddony; then pot them, and pour the syrup into them before they be cold.

Take your Apricocks and Pear-Plums and boil them tender, then take as much Sugar as they do weigh, and take as much water as will make the syrup, take your green Peaches before they be stoned and thrust a pin through them, and then make a strong water of ashes, and cast them into the hot standing lye to take off the fur from them, then wash them in three or four waters warm, so then put them into so much clarified Sugar as will candy them; so boil them, and put them up, &c.

To dry Pippins, or Pears without Sugar.

Take Pippins or Pears and prick them full of holes with a bodkin, & lay them in sweet wort three or four dayes, then lay them on a sieves bottom, till they be dry in an Oven, but a drying heat. This you may do to any tender Plum.

To make Syrup of Clove-gilly flowers.

Take a quart of water, half a bushel of Flowers, cut off the whites, and with a sieve sift away the seeds, bruise them a little; let your water be boiled, and a little cold again, then put in your Flowers, and let them stand close covered twenty four hours; you may put in but half the flowers at a time, the strength will come out the better;

to that liquor put in four pound of Sugar, let it lye in all night, next day boil it in a Gallipot, set it in a pot of water, and there let it boil till all the Sugar be melted and the syrup be pretty thick, then take it out, and let it stand in that till it be through cold, then glass it.

To make Syrup of Hysop for Colds.

Take a handful of Hysop, of Figs, Raisins, Dates, of each an ounce, of Collipint half an handful, French Barley one ounce, boil therein three pints of fair water to a quart, strain it and clarifie it with two whites of Eggs, then put in two pound of fine sugar, and boil it to a syrup.

To make Orange Water.

Take a pottle of the best Maligo Sack, and put in as many of the peels of Oranges as will go in, cut the white clean off, let them steep twenty four hours; still them in a glass still, and let the water run into the Receiver upon fine Sugar-candy; you may still it in an ordinary Still.

To dry Cherries.

Take a pound of sugar, dissolve it in thin fair water, when it is boiled a little while, put in your Cherries after they are stoned, four pound to one pound of Sugar, let them lye in the Sugar three dayes, then take them out of the syrup and lay them on sieves one by one, and set them before the Sun upon stools, turn them every day, else they will mould; when they look of a dark red colour, and are dry then put them up. And so you may do any manner of Fruit. In the Sun is the best drying of them, put into the syrup some juyce of Rasps.

To make juyce of Liquorish.

Take English Liquorish, and stamp it very clean, bruise it with a hammer, and cut it in peices; to a pound of Liquorish thus bruised, put a quart of Hysop water, let them soak together in an earthen pot a day and a night, then pull the Liquorish into small pieces, and lay it in soak again two dayes more; then strain out the Liquorish, and boil the liquor a good while. Stir it often; then put in half a pound of Sugar-candy, or Loaf-sugar finely beaten, four grains of Musk, as much Ambergreece, bruise them small with a little Sugar; then boil them together till it be good & thick, still have care you burn it not; then put it out in glass plates, and make it into round rolls, and set it in a drying place till it be stiff, that you may work it into rolls to be cut as big as Barley corns, and so lay them on a place again: If it be needful strew on the place again a little Sugar to prevent thickning; so dry them still if there be need and if they should be too dry, the heat of the fire will soften them again.

A Perfume for Cloths, Gloves.

Take of Linet two grains, of Musk three, of Ambergreece four, and the oyl of Bems a pretty quantity; grinde them all upon a Marble stone fit for that purpose; then with a brush or sponge rake them over, and it will sweeten them very well; your Gloves or Jerkins must first be washed in red Rose-water, and when they are almost dry, stretch them forth smooth, and lay on the Perfumes.

To make Almond Bisket.

Take the whites of four new laid Eggs, and two yolks, then beat it well for an hour together, then have in readiness a quarter of a pound of the best Almonds blanched in cold water, & beat them very small with Bose-wart, for fear of Oyling; then, have a pound of the best Loaf-sugar finely beaten, beat that in the Eggs a while, then put in your Almonds, and five or six spoonfuls of the finest flower, and so bake them together upon Paper plates, you may have a little fine Sugar in a piece of tiffany to dust them over as they be in the Oven, so bake them as you do Bisket.

To dry Apricocks.

First stone them, then weigh them, take the weight of them in double refined Sugar, make the syrup with so much water as will wet them, and boil it up so high, that a drop being droped on a Plate it will slip clean off, when it is cold, put in your Apricocks being pared, whilst your Syrup is hot, but it must not be taken off the fire before you put them in, then turn them in the syrup often, then let them stand 3 quarters of an hour, then take them out of the syrup, and tie them up in Tiffanies, one in a tiffany or more, as they be in bigness, and whilst you are tying them up, set the syrup on the fire to heat, but not to boil, then put your Apricocks into the syrup, and set them on a quick fire, and let them boil, as fast as you can, skim them clean, and when they look clear take them from the fire, and let them lie in the syrup till the next day, then set them on the fire to heat, but not to boil; then set them by till the next day, and lay them upon a clean Sieve to drain, and when they are well drained, take them out of the Tiffanies, and so dry them in a Stove, or better in the Sun with Glasses over them, to keep them from the dust.

To make Quinces for Pies.

Wipe the Quinces, and put them into a little vessel of swall Beer when it hath done working; stop them close that no air can get in, and this will keep them fair all the year and good.

The best way to break sweet Powder.

Take of Orrice one pound, Calamus a quarter of a pound, Benjamin one half pound, Storax half a pound, Civet a quarter of an ounce, Cloves a quarter of a pound, Musk one half ounce, Oyl of Orange flowers one ounce, Lignum Aloes one ounce, Rosewood a quarter of a pound, Ambergreece a quarter of an ounces. To every pound of Roses put a pound of powder; the bag must be of Taffity, or else the powder will run through.

To make excellent Perfumes.

Take a quarter of a pound of Damask Rose-buds cut clean from the Whites, stamp them very small, put to them a good spoonful of Damask Rose-water, so let them stand close stoopped all night, then take one ounce and a quarter of Benjamin finely beaten, and also searsed, (if you will) twenty grains of Civit, and ten grains of Musk; mingle them well together, then make it up in little Cakes between Rose leaves, and dry them between sheets of Paper.

To make Conserve of Roses boiled.

Take a quart of red Rose-water, a quart of fair water, boil in the Water a pound of red Rose-leaves, the whites cut off, the leaves must be boiled very tender; then take three pound of Sugar, and put to it a pound at a time, and let it boil a little between every pound, so put it up in your pots.

To make Conserves of Roses unboiled.

Take a pound of red Rose leaves, the whites cut off, stamp them very fine, take a pound of Sugar, and beat in with the Roses, and put it in a pot, and cover it with leather, and set it in a cool place.

To make a very good Pomatum.

Take the Fat of a young Dog one pound, it must be killed well that the blood settle not into the fat, then let the outer skin be taken off before it be opened, lest any of the hair come to the fat, then take all the fat from the inside, and as soon as you take it off fling it into Conduit water, and if you see the second skin be clear, peel it and water it with the other: be sure it cools not out of the water: you must not let any of the flesh remain on it, for then the Pomatum will not keep. To one pound of this fat take two pound of Lambs caule, and put it to the other in the water and when you see it is cold, drain it from the water in a Napkin, and break it in little peices with your fingers, and take out all the little veins; then take eight ounces of Oyl of Tartar, and put in that first, stiring it well together, then put it into a Gallon of Conduit water, and let it stand till night; shift this with so much Oyl and Water, morning and evening seven dayes together, and be sure you shift it constantly; and the day before you mean to melt it wring it hard by a little at a time, and be sure the Oyl and water be all out of it, wring the water well out of it with a Napkin every time you shift it; then put in three pints of Rose-water; let it stand close covered twelve hours, then wring out that, and put it in a pint of fresh Rose-water into a high Gallipot with the *Fæces*; then tie it close up, and set it in a pot of water, and let it boil two hours then take it out, and strain it into an earthen Pan, let it stand till it be cold; then cut a hole in it, and let out the water, then scrape away the bottom, and dry it with a cloth, and dry the pan, melt it in a Chafing-dish of Coales, or in the Gallipots; beat it so long till it look very white and shining; then with your hand fling it in fine Cakes upon white paper, and let it lye till it be cold, then put it into Gallipots. This will be very good for two or three years.

To make Raisin Wine.

Take two pound of Raisins of the Sun shred, a pound of good powdered Sugar, the juice of two Lemons, one pill, put these into an earthen Pot with a top, then take two gallons of water, let it boil half an hour, then take it hot from the fire, and put it into the pot, and cover it close for three or four dayes, stirring it twice a day, being strained put it into bottles, and stop it more close, in a fortnight or three weeks it may be drunk; you may put in Clove Gilly flowers, or Cowslips, as the time of the year is when you make it; and when you have drawn this from the Raisins, and bottled it up, heat two quarts of water more, put it to the ingredients, and let it stand as aforesaid. This will be good, but smaller than the other, the water must be boiled as the other.

To make Rasberry Wine.

Take a Gallon of good Rhenish Wine, put into it as much Rasberries very ripe as will make it strong, put it in an earthen pot, and let it stand two dayes, then pour your Wine from your Rasberries, and put into every bottle two ounces of Sugar, stop it up and keep it by you.

The best way to preserve Cherries.

Take the best Cherries you can get, and cut the stalks something short, then for every pound of these Cherries take two pound of other Cherries, and put them of their stalks and stones, put to them ten spoonfuls of fair water, and then set them on the fire to boil very fast till you see that the colour of the syrup be like pale Claret wine, then take it off the fire, and drain them from the Cherries into a Pan to preserve in. Take to every pound of Cherries a quarter of Sugar, of which take half, and dissolve it with the Cherry water drained from the Cherries, and keep them boiling very fast till they will gelly in a spoon, and as you see the syrup thin, take off the Sugar that you kept finely beaten, and put it to the Cherries in the boiling, the faster they boil, the better they will be preserved, and let them stand in a Pan till they be almost cold.

A Tincture of Ambergreece.

Take Ambergreece one ounce, Musk two drams, spirit of Wine half a pint, or as much as will cover the ingredients two or three fingers breadth, put all into a glass, stop it close with a Cork and Bladder; set it in Horse dung ten or twelve days, then pour off gently the Spirit of Wine, and keep it in a Glass close stopt, then put more spirit of Wine on the Ambergreece, and do as before, then pour it off, after all this the Ambergreece will serve for ordinary uses. A drop of this will perfume any thing, and in Cordials it is very good.

To make Usquebath the best way.

Take two quarts of the best *Aqua vitæ*, four ounces of scraped liquorish, and half a pound of sliced Raisins of the Sun, Anniseeds four ounces, Dates and Figs, of each half a pound, sliced Nutmeg, Cinnamon, Ginger, of each half an ounce, put these to the *Aqua vitæ*, stop it very close, and set it in a cold place ten dayes, stirring it twice a day with a stick, then strain and sweeten it with Sugar-candy; after it is strained, let it stand till it be clear, then put into the glass Musk and Ambergreece; two grains is sufficient for this quantity.

To preserve Cherries with a quarter of their weights in Sugar.

Take four pound of Cherries, one pound of Sugar, beat your Sugar and strew a little in the bottom of your skillet, then pull off the stalk and stones of your Cherries, and cut them cross the bottom with a knife; let the juyce of the Cherries run upon the Sugar; for there must be no other liquor but the juyce of the Cherries; cover your Cherries over with one half of your Sugar, boil them very quick, when they are half boiled, put in the remainder of your sugar, when they are almost enough, put in the rest of the sugar; you must let them boil till they part in sunder like Marmalade, stirring them continually; so put them up hot into your Marmalade glasses.

To make Gelly of Pippins.

Take Pippins, and pare them, and quarter them, and put as much water to them as will cover them, and let them boil till all the vertue of the Pippins are out; then strain them, and take to a pint of that liquor a pound of Sugar, and cut long threads of Orange peels, and boil in it, then take a Lemon, and pare and slice it very thin, and boil it in your liquor a little thin, take them out, and lay them in the bottom of your glass, and when it is boiled to a gelly, pour it on the Lemons in the glass. You must boil the Oranges in two or three waters before you boil it in the gelly.

To make Apricock Cakes.

Take the fairest Apricocks you can get, and parboil them very tender, then take off the Pulp and their weight of Sugar, and boil the Sugar and Apricocks together very fast, stir them ever lest they burn to, and when you can see the bottom of the Skillet it is enough; then put then into Cards sowed round, and dust them with fine Sugar, and when they are cold stone them, then turn them, and fill them up with some more of the same stuff; but you must let them stand for three or four dayes before you turn them off the first place; and when you find they begin to candy, take them out of the Cards, dust them with Sugar again; so do ever when you turn them.

To preserve Barberries the best way.

First stone them and weigh them, half a pound of sugar to half a pound of them, then pair them and slice them into that liquor, take the weight of it in sugar; then take as many Rasberries as will colour it, and strain them into the liquor, then put in the sugar, boil it as fast as you can, then skim it till it be very clear, then put in your Barberries, and that sugar you weighed, and so let them boil till the skin be fully risen up, then take them off, and skin them very clean, and put them up.

To make Lozenges of Red Roses.

Boil your sugar to sugar again, then put in your Red Roses being finely beaten and made moist with the juyce of a Lemmon, let it not boil after the Roses are in but pour it upon a Pye-plate, and cut it into what form you please.

To make Chips of Quinces.

First scald them very well, then slice them into a Dish, and pour a Candy Syrup to them scalding hot, and let them stand all night, then lay them on plates, and searse sugar on them, and turn them every day, and scrape more sugar on them till they be dry. If you would have them look clear, heat them in syrup, but not to boil.

To make Sugar of Wormwood, Mint, Anniseed, or any other of that kinde.

Take double refined Sugar, and do but wet it in fair water, or Rose-water and boil it to a Candy, when it is almost boiled take it off, and stir it till it be cold; then drop in three or four drops of the Oyls of whatsoever you will make, and stir it well; then drop it on a board, being before fitted with Sugar.

To make Syrup of Lemons or Citrons.

Pare off all the rindes, then slice your Lemmons very thin, and lay a lare of Sugar finely beaten, and a lare of Lemons in a silver Bason till you have filled it, or as much as you mean to make, & so let it stand all night; the next day pour off the liquor that runs from it into a glass through a Tiffany strainer. Be sure you put sugar enough to them at the first, and it will keep a year good, if it be set up well.

To make Jambals of Apricocks or Quinces.

Take Apricocks or Quinces, and quoddle them tender, then take their Pulp and dry it in a dish over a Chafing-dish of coals, and set it in a Stove for a day or two; then beat it in a stone Mortar, putting in as much Sugar as will make a stiff paste; then colour it with Saunders, Cochinele or blew Starch, and make it up in what colour you please, rowl them with battle doors into long pieces, and tye them up in knots, and so dry them.

To make Cherry-water.

Take nine pound of Cherries, pull out the stones and stalks, break them with you hand, and put them into nine pints of Claret Wine, take nine ounces of Cinamon, and three Nutmegs, bruise them, and put them into this, then take of Rosemary and Balm, of each half a handful, of sweet Marjoram a quarter of a handful; put all these with the aforenamed into an earthen pot well leaded; so let them stand to infuse twenty four hours; so distil it in a Limbeck, keeping the strongest water by it self, put some sugar finely beaten into your glasses. If your first water be too strong, put some of the second to it as you use it. If you please you may tye some Musk and Ambergreese, in a rag, and hang it by a thread in your glass.

To make Orange Cakes.

Take Oranges and pare them as thin as you can, then take out the meats clean, and put them in water; let them lye about an hour, shift the water, and boil them very tender in three or four waters, then put them up, and dry them on a cloath: mince them as small as you can, then put them into a dish, and squeeze all the juyce of the meat into them, and let them stand till the next day, take to every pound of these a pound and a quarter of double refined Sugar. Boil it with a spoonful of water at the bottom to keep it from burning till it be Sugar again; then put in your Oranges and let them stand and dry on the fire, but not boil; then put them on glass plates, and put them in a stove, the next day make them into Cakes, and so fry them as fast as you can.

To preserve Oranges the French way.

Take twelve of the fairest Oranges and best coloured, and if you can get them with smooth skins they are the better, and lay them in Conduit water, six dayes and nights, shifting them into fresh water morning and evening; then boil them very tender, and with a knife pare them very thin, rub them with salt, when you have so done, core them with a coring Iron, taking out the meat and seeds; then rub them with a dry cloth till they be clean, add to every pound of Oranges a pound and half of Sugar, and to a pound of sugar a pint of water; then mingle your, sugar and water well together in a large skillet or pan; beat the whites of three Eggs and put that into it, then set it on the fire, and let it boil till it rises, and strain it through a Napkin; then set it on the fire again, and let it boil till the syrup be thick, then put in your Oranges, and make them seethe as fast as you can, now and then putting in a piece of fine loaf Sugar the bigness of a Walnut, when they have boiled near an hour, put into them a pint of Apple water; then boil them apace, and add half a pint of white Wine, this should be put in before the Apple-water, when your Oranges are very clear, & your Syrup is so thick that it will gelly, (which you may know by setting some to cool in a spoon) when they are ready to be taken off from the fire; then put in the juyce of eight Lemons warm into them, then put them into an earthen pan, and so let them stand till they be cold, then put every Orange in a several glass or pot; if you do but six Oranges at a time it is the better.

To preserve green Plums.

The greatest Wheaten Plum is the best, which will be ripe in the midst of *July*, gather them about that time, or later, as they grow in bigness, but you must not suffer them to turn yellow, for then they never be of good colour; being gathered, lay them in water for the space of twelve hours, and when you gather them, wipe them with a clean linnen cloth, and cut off a little of the stalks of every one, then set two skillets of water on the fire, and when one is scalding hot put in your Plums, and take them from the fire, and cover them, and let them rest for the space of a quarter of an hour; then take them up, and when your other skillet of water doth boil, put them into it; let them but stay in it a very little while, and so let the other skillet of water, wherein they were first boiled, be set to the fire again, and make it to boil, and put in your Plums as before, and then you shall see them rivet over, and yet your Plums very whole; then while they be hot, you must with your knife scrape away the riveting; then take to every pound of Plums a pound and two ounces of Sugar finely beaten, then set a pan with a little fair water on the fire, and when it boils, put in your Plums, and let them settle half a quarter of an hour till you see the colour wax green, then set them off the fire a quarter of an hour, and take a handful of Sugar that is weighed, and strow it in the bottom of the pan wherein you will preserve, and so put in your Plums one by one, drawing the liquor from them, and cast the rest of your Sugar on them; then set the pan on a moderate fire, letting them boil continually but very softly, and in three quarters of an hour they will be ready, as you may perceive by the greenness of your Plums, and thickness of your syrup, which if they be boiled enough, will gelly when it is cold; then take up your Plums, and put them into a Gallipot, but boil your Syrup a little longer, then strain it into some vessel, and being blood-warm, pour it upon your plums, but stop not the pot before they be cold. Note also you must preserve them in such a pan, as they may lye one by another, and turn of themselves; and when they have been five or six days in the syrup, that the syrup grow thin, you may boil it again with a little Sugar, but put it not to your Plums till they be cold. They must have three scaldings, and one boiling.

To dry Plums.

Take three quarters of a pound of Sugar to a pound of black Pear-plums, or Damsins, slit the Plums in the crest, lay a lay of Sugar with a lay of Plums, and let them stand all night; if you stone the Plums, fill up the place with sugar, then boil them gently till they be very tender, without breaking the skins, take them into an earthen or silver dish, and boil your syrup afterwards for a gelly, then pour it on your Plums scalding hot, and let them stand two or three dayes, then let them be put to the Oven after you draw your bread, so often untill your syrup be dryed up, and when you think they are almost dry, lay them in a sieve, and pour some scalding water on them, which will run through the sieve, and set them in an Oven afterwards to dry.

To preserve Cherries the best way, bigger than they grow naturally, &c.

Take a pound of the smallest Cherries, and boil them tender in a pint of fair water, then strain the liquor from the substance, then take two pound of good Cherries, and put them into a preserving-pan with a lay of Cherries, and a lay of sugar: then pour the syrup of the other Cherries about them, and so let them boil as fast as you can with a quick fire, that the syrup may boil over them, and when your syrup is thick and of good colour, then take them up, and let them stand a cooling by partitions one from another, and being cold you may pot them up.

To preserve Damsins, red Plums or black.

Take your Plums newly gathered, and take a little more sugar than they do weigh, then put to it as much water as will cover them; then boil your syrup a little while, and so let it cool, then put in your Damsins or Plums, then boil them leasurely in a pot of seething water till they be tender, then being almost cold pot them up.

To dry Pippins or Pears.

Take your Pippins, Pears, Apricocks, pare them, and lay them in a broad earthen pan one by one, and so rowl them in searsed Sugar as you flower fried fish; put them in an Oven as hot as for manchet, and so take them out, and turn them as long as the Oven is hot; when the Oven is of a drying heat, lay them upon a Paper, and dry them on the bottom of a Sieve; so you may do the least Plum that is.

To dry Pippins or Pears another way.

Take Pippins or Pears, and lay them in an earthen Pan one by one, and when they be baked plump and not broken, then take them out, and lay them upon a Paper, then lay them on a Sieves bottom, and dry them as you did before.

To dry Apricocks tender.

Take the ripest of the Apricoks, pare them, put them into a silver or earthen skillet, and to a pound of Apricocks put three quarters of a pound of Sugar, set your Apricocks over your fire; stirring them till they come to a pulp, and set the Sugar in another skillet by boiling it up to a good height, then take all the Apricocks, and stir them round till they be well mingled, then let it stand till it be something cold and thick, then put it into cards, being cut of the fashion of an Apricock, and laid upon glass plates; fill the Cards half full, then set them in your stove, but when you find they are so dry that they are ready to turn, then provide as much of your pulp as you had before, and so put to every one a stove, when they are turned, (which you must have laid before) & pour the rest of the Pulp upon them, so set them into your stove, turning them till they be dry.

To dry Plums.

Take a pound of Sugar to a pound of Plums, pare them, scald your Plums, then lay your Plums upon a sieve till the water be drained from them, boil your Sugar to a Candy height, and then put your Plums in whilst your syrup is hot, so warm them every morning for a week, then take them out, and put them into your stove and dry them.

To dry Apricocks.

Take your Apricocks, pare and stone them, then weigh half a pound of sugar to a pound of Apricocks, then take half that sugar, and make a thin syrup, and when it boileth, put in the Apricocks; then scald them in that syrup; then take them off the fire, and let them stand all night in that syrup, in the morning take them out of that syrup, and make another syrup with the other half of the sugar, then put them in, and preserve them till they look clear; but be sure you do not do them so much as those you keep preserved without drying; then take them out of that syrup, and lay them on a piece of Plate till they be cold; then take a skillet of fair water, and when the water boils take your Apricocks one after another in a spoon, and dip them in the water first on one side, and then on the other; not letting them go out of the spoon: you must do it very quick, then put them on a piece of plate, and dry them in a Stove, turning them every day; you must be sure that your Stove or Cupboard where you dry them, the heat of it be renewed three times a day with a temperate drying heat untill they be something dry, then afterwards turn once as you see cause.

Conserves of Violets the Italian manner.

Take the leaves of blue Violets separated from their stalks and greens, beat them very well in a stone Mortar, with twice their weight of Sugar, and reserve them for your use in a glass vessel.

The Vertue.

The heat of Choller it doth mitigate extinguisheth thirst, asswageth the belly, and helpeth the Throat of hot hurts, sharp droppings and driness, and procureth rest: It will keep one year.

Conserves of red Roses the Italian manner.

Take fresh red Roses not quite ripe, beat them in a stone Mortar, mix them with double their weight of Sugar, and put them in a glass close stopped, being not full, let them remain before you use them three months, stirring of them once a day.

The Vertues.

The Stomach, Heart, and Bowels it cooleth, and hindreth vapours, the spitting of blood and corruption for the most part (being cold) it helpeth. It will keep many years.

Conserve of Borage Flowers after the Italian manner.

Take fresh Borage flowers cleansed well from their heads four ounces, fine sugar twelve ounces, beat them well together in a stone Mortar, and keep them in a vessel well placed.

The vertues are the same with Bugloss flowers.

Conserve of Rosemary flowers after the Italian manner.

Take new Rosemary Flowers one pound, of white sugar one pound; so beat them together in a Marble Mortar with a wooden Pestle, keep it in a gallipot, or vessel of earth well glassed, or in one of hard stone. It may be preserved for one year or two.

The Vertues.

It comforteth the heart, the stomach, the brain, and all the nervous part of the Body.

Conserve of Betony after the Italian way.

Betony new and tender one pound, the best sugar three pound, beat them very small in a stone Mortar, let the sugar be boiled with two pound of Betony-water to the consistance of a syrup, at length mix them together by little and little over a small fire, and make a Conserve, which keep in a glass.

The Vertues.

It helpeth the cold pains of the head, purgeth the stomach and womb: it helpeth stoniness of the Reins, and furthereth Conception.

Conserve of Sage.

Take new flowers of Sage one pound, sugar one pound; so beat them together very small in a Marble Mortar, put them in a vessel well glassed and steeped, set them in the Sun, stir them daily; it will last one year.

The Vertues.

It is good in all cold hurts of the brain, it refresheth the Stomach, it openeth obstructions and takes away superfluous and hurtfull humours from the stomach.

Conserve of flowers of Lavender.

Take the flowers being new, so many as you please, and beat them with three times their weight of white Sugar, after the same manner as Rosemary flowers; they will keep one year.

The Vertues.

The Brain, the Stomach, Liver, Spleen, and Womb it maketh warm, and is good in the Suffocation of the Womb, hardness of the spleen and for the Apoplexy.

Conserve of Marjoram.

The Conserve is prepared as Betony, it keepeth a year.

The Vertues.

It is good against the coldness, moistness of the Brain, and Stomach, and it strengthneth the Vital spirits.

Conserve of Peony after the Italian way.

In the Spring take of the Flowers fresh half a pound, Sugar one pound, beat them together in a good stone Mortar, then put them in a glass, and set them in the sun for three months, stirring them daily with a wooden Spathula.

The Vertues.

It is good against the Falling-sickness, and giddiness in the head, it cleanseth the Reins and Bladder.

Touching Candies, as followeth.

To Candy Rosemary-flowers in the Sun.

Take Gum-Dragon, and steep it in Rose-water, then take the Rosemary flowers, good coloured, and well pickt, and wet them in the water that your Gum dragon is steeped in, then take them out, and lay them upon a paper, and strew fine Sugar over them; this do in the hot sun, turning them, and strewing Sugar on them, till they are candied, and so keep them for your use.

To Make Sugar of Roses.

Take the deepest coloured red Roses, pick them, cut off the white bottoms, and dry your red leaves in an Oven, till they be as dry as possible, then beat them to powder and searse them, then take half a pound of Sugar beaten fine, put it into your pan with as much fair water as will wet it; then set it in a chaffing-dish of coals, and let it boil till it be sugar again, then put as much powder of Roses as will make it look very red stir them well together, and when it is almost cold, put it into pailes, and when it is throughly cold, take them off, and put them in boxes.

To Candy Pippins, Pears, Apricocks or Plums.

Take of these fruits being pared, and strew sugar upon them, as you do flower upon frying fish; then lay them on a board in a Pewter dish, so put them into an Oven as hot as for Manchet; as the liquor comes from them, pour forth, turn them, and strew more Sugar on them, and sprinkle Rose-water on them, thus turning and sugaring of them three or four times, till they be almost dry, then lay them on a Lettice Wire, or on the bottom of a sieve in a warm Oven, after the bread is drawn out, till they be full dry: so you may keep them all the year.

To Candy or clear Rockcandy flowers.

Take spices, and boil them in a syrup of Sugar, then put in the flowers, boil them till they be stiff, when you spread them on a Paper, lay them on round Wires in an earthen pan, then take as much hard Sugar as will fill your pan, and as much water as will melt the sugar, that is half a pint to every pound; then beat a dozen spoonfuls of fair water, and the white of an Egg in a bason, with a birchen rod till it come to a Froth, when your sugar is melted and boiled, put the froth of the Egg in the hot syrup, and as it riseth, drop in a little cold water; so let it boil a little while, then scum it, then boil it to a Candy height, that is, when you may draw it in small threads between your finger and your thumb: then pour forth all your syrup that will run from it in your pan, then set it a drying one hour or two, which done pick up the wiers, and take off the flowers, and lay them on papers, and so dry them.

To Candy Spanish Flowers.

Take the Blossoms of divers sorts of flowers, and make a syrup of water and sugar, and boil it very thick, then put in your Blossoms, and stir them in their boiling, till it turn to sugar again, then stir them with the back of a spoon, till the Sugar fall from it; so may you keep them for Sallets all the year.

To Candy Grapes, Cherries or Barberries.

Take of these fruits, and strew fine sifted sugar on them, as you do flower on frying fish, lay them on a lattice of wier in a deep earthen pan, and put them into an Oven as hot as for Manchet; then take them out, and turn them and sugar them again, and sprinkle a little Rose-water on them, pour the syrup forth as it comes from them, thus turning and sugaring them till they be almost dry, then take them out of the earthen pan, and lay them on a lattice of wire, upon two billets of wood in a warm Oven, after the bread is drawn, till they be dry and well candied.

To Candy Suckets of Oranges, Lemons, Citrons, and Angelica.

Take, and boil them in fair water tender, and shift them in three boilings, six or seven times, to take away their bitterness, then put them into as much Sugar as will cover them, and so let them boil a walm or two, then take them out, and dry them in a warm Oven as hot as Manchet, and being dry boil the Sugar to a Candy height, and so cast your Oranges into the hot Sugar, and take them out again suddenly, and then lay them upon a lattice of Wyer or the bottom of a Sieve in a warm Oven after the bread is drawn, still warming the Oven till it be dry, and they will be well candied.

To Candy the Orange Roots.

Take the Orange Roots being well and tenderly boiled, petch them and peel them, and wash them out of two or three waters; then dry them well with a fair cloth; then pot them together two or three in a knot, then put them into as much clarified Sugar as will cover them, and so let them boil leisurely, turning them well until you see the Sugar drunk up into the Root; then shake them in the Bason to sunder the knits; and when they wax dry, take them up suddenly, and lay them on sheets of white Paper, and so dry them before the fire an hour or two, and they will be candied.

Candy Orange Peels after the Italian way.

Take Orange Peels so often steeped in cold water, as you think convenient for their bitterness, then dry them gently, and candy them with some convenient syrup made with Sugar, some that are more grown, take away that spongious white under the yellow peels, others do both together.

The Vertues.

They corroborate the Stomach and Heart.

To Candy Citrons after the Spanish way.

Take Citron Peels so large as you please the inner part being taken away, let them be steeped in a clear lye of water and ashes for nine dayes, and shift them the fifth day, afterward wash them in fair water, till the bitterness be taken away, and that they grow sweet, then let them be boiled in fair water till they grow soft, the watry part being taken away, let them be steeped in a vessel of stone twenty four hours, with a Julip, made of white Sugar and three parts water; after let them be boiled upon a gentle fire, to candiness of Penidies or Paste; being taken out of that, let them be put into a glass vessel, one by one, with the julip of Roses made somewhat hard or with sugar; some do add Amber and Musk to them.

The Vertues.

It comforteth the Stomach and Heart, it helpeth concoction.

Candied Cherries, the Italian way.

Take Cherries before they are full ripe, the stones taken out, put clarified sugar boiled to a height, then pour it on them.

Chicory Roots candied the Italian way.

Take Chicory new and green, the outward Bark being taken away, then before they be candied, let them be cut in several parts, and gently boiled, that no bitterness may remain, then set them in the air placed severally, and put sugar to them boiled to a height.

Touching Marmalets, and Quiddony, as followeth.

To make Marmalet of Damsins.

Take two quarts of Damsins that be through ripe, and pare off the skin of three pints of them, then put them into an earthen Pipkin, those with the skins undermost then set the Pipkin into a pot of seething water, and let the water seethe apace untill the Damsins be tender. Cover the Pipkin close, that no water gets into them, and when they are tender, put them out into an earthen pan, and take out all the stones and skins, and weigh them, and take the weight with hard sugar, then break the sugar fine, and put it into the Damsins, then set it on the fire, and make it boil apace till it will come from the bottome of the skillet, then take it up, and put it into a glass but scum it clear in the boiling.

To make white Marmalet of Quinces.

Take unpared Quinces, and boil them whole in fair water, peel them and take all the pap from the core, to every pound thereof add three quarters of a pound of Sugar, boil it well till it comes well from the pans bottom, then put it into boxes.

To make Marmalet of any tender Plum.

Take your Plums, & boil them between two dishes on a Chafing dish of coals, then strain it, and take as much Sugar as the Pulp doth weigh, and put to it as much Rose-water, and fair water as will melt it, that is, half a pint of water to a pound of Sugar, and so boil it to a Candy height, then put the pulp into hot sugar, with the pap of a roasted apple. In like manner you must put roasted apples to make Past Royal of it, or else it will be tough in the drying.

To make Orange Marmalet.

Take Oranges, pare them as thin as you can; boil them in four several waters, let them be very soft before you take them out, then take two quarts of Spring-water, put thereto twenty Pippins pared, quartered, and coared, let them boil till all the vertue be out, take heed they do not lose the colour; then strain them, put to every pint of water a pound of sugar, boil it almost to a Candy-height, then take out all the meat out of the Oranges, slice the peel in long slits as thin as you can, then put in your peel with the juyce of two Lemmons, and one half Orange, then boil it to a Candy.

To make Quiddony of Pippins of Ruby or any Amber colour.

Take Pippins, and cut them in quarters, and pare them, and boil them with as much fair water as will cover them, till they be tender, and sunk into the water, then strain all the liquor from the Pulp, then take a pint of that liquor, and half a pound of Sugar, and boil it till it be a quaking gelly on the back of a spoon; so then pour it on your moulds, being taken out of fair water; then being cold turn them on a wet trencher, and so slide them into the boxes, and if you would have it ruddy colour, then boil it leasurely close covered, till it be as red as Claret Wine, so may you conceive, the difference is in the boiling of it; remember to boil your Quinces in Apple-water as you do your Plums.

To make Quiddony of all kind of Plums.

Take your Apple-water, and boil the Plums in it till it be red as Claret Wine, and when you have made it strong of the Plums, put to every pint half a pound of Sugar, and so boil it till a drop of it hang on the back of a spoon like a quaking gelly. If you will have it of an Amber colour, then boil it with a quick fire, that is all the difference of the colouring of it.

To make Marmalet of Oranges, or Orange Cakes, &c.

Take the yellowest and fairest Oranges, and water them three days, shifting the water twice a day, pare them as thin as you possible can, boil them in a water changed five or six times, until the bitterness of the Orange be boiled out, those that you preserve must be cut in halves, but those for Marmalet must be boiled whole, let them be very tender, and slice them very thin on a Trencher, taking out the seeds and long strings, and with a Knife make it as fine as the Pap of an Apple; then weigh your Pap of Oranges, and to a pound of it, take a pound and a half of sugar; then you must have Pippins boiled ready in a skillet of fair water, and take the pap of them made fine on a Trencher, and the strings taken out, (but take not half so much Pippins as Oranges) then take the weight of it in sugar, and mix it both together in a Silver or Earthen Dish; and set it on the coals to dry the water out of it, (as you do with Quince Marmalet) when your sugar is Candy height, put in your stuff, and boil it till you think it stiff enough, stirring it continually: if you please you may put a little Musk in it.

Touching Pastrey and Pasties.

To make Sugar Cakes.

Take three pound of the finest Wheat Flower, one pound of fine Sugar, Cloves, and Mace of each one ounce finely searsed, two pound of butter, a little Rose-water, knead and mould this very well together, melt your butter as you put it in; then mould it with your hand forth upon a board, cut them round with a glass, then lay them on papers, and set them in an Oven, be sure your Oven be not too hot, so let them stand till they be coloured enough.

To make clear Cakes of Plums.

Take Plums of any sorts, Raspiss are the best, put them in a stone Jug, into a pot of seething water, and when they are dissolved, strain them together through a fair cloth, and take to a pint of that a pound of sugar, put to as much color as will melt it, and boil to a Candy height; boil the liquor likewise in another Posnet, then put them seething hot together, and so boil a little while stirring them together, then put them into glasses, and set them in an Oven or Stove in a drying heat, let them stand so two or three weeks, and never be cold, removing them from one warm place to another, they will turn in a week; beware you set them not too hot, for they will be tough; so every day turn them till they be dry; they will be very clear.

To make Paste of Oranges and Lemons.

Take your Oranges well coloured, boil them tender in water, changing them six or seven times in the boiling, put into the first water one handful of Salt, and then beat them in a wooden bowl with a wooden Pestle, and then strain them through a piece of Cushion Canvas, then take somewhat more than the weight of them in Sugar, then boil it, dry and fashion it as you please.

To make Rasberry Cakes.

Take Rasberries, and put them into a Gallipot, cover them close, and set them into a skillet of water, and let them boil till they are all to mash, then rub them through a strainer of Cushion Canvas, put the liquor into a silver bason, and set it upon a very quick fire; and put into it one handful or two of whole Rasberries, according to the quantity of your liquor; and as you shall like to have seeds in your paste: Thus let it boyl very fast till it be thick; and continually stir, lest it burn; then take two silver dishes that are of a weight, and put them into your scales, in the one put the Raspiss stuffe, and in the other double refined Sugar finely beaten, as much as the weight of Raspiss stuff; then put as much water to the sugar as will melt it, set it upon the fire, and let it boil till it be very high candied, then take it from the fire, and put your Raspiss stuff into it; and when your Sugar and Rasberries are very well mixt together, and the sugar well melted from about the dish, (which if it will not do from the fire, set it on again) but let it not boil in any case; when it is pretty cool, lay it by spoonfuls in places, and put it into your stuff, keeping temperate fire to it twice a day till it be candied that will turn them, joyn two of the pieces together, to make the cakes the thicker.

To make Paste of Genoa Citrons.

Take Citrons, & boil them in their skins, then scrape all the pulp from the core, strain it through a piece of Cushion Canvas, take twice the weight of the pulp in Sugar, put to it twice as much water as will melt it that is half a pint to every pound of Sugar, boil it to a Candy height; dry the Pulp upon a Chafing-dish of Coales, then put the syrup and the Pulp hot together, boil it with stirring until it will lye upon a Pye-plate, set it in a warm stone Oven upon two billets of wood, from the heat of the Oven, all one night, in the morning turn it, and set it in the like heat again, so turn it every day till it be dry.

To make a French Tart.

Take a quarter of Almonds or thereabouts, and peel them, then beat them in a mortar, take the white of the breast of a cold Capon, and take so much Lard as twice the quantity of the Capon, and so much Butter, or rather more, and half a Marrow-bone, and if the bone be little then all the Marrow, with the juyce of one Lemon; beat them all together in a Mortar very well, then put in one half pound of loaf sugar grated, then take a good piece of Citron, cut it in small pieces, and half a quarter of Pistanius, mingle all these together, take some flour, and the yolks of two or three Eggs, and some sweet Butter, and work it with cold water.

To make Cakes of Pear Plums.

Take a pound of the clear, or the Pulp, a pound of Sugar, and boil it to a Sugar again, then break it as small as you can, and put in the clear, when your Sugar is melted in it, and almost cold, put it in glass plates, and set them into your stove as fast as you can, with coals under them, and so twice a day whilst they be dry enough to cut; if you make them of the clear, you must make paste of Apples to lay upon them, you must scald them, and beat them very well, and so use them as you do your Plums, and then you may put them into what fashion you please.

To make Cakes, viz.

Take a pound of Sugar finely beaten, four yolks of Eggs, two whites, one half pound of Butter washt in Rose-water, six spoonfuls of sweet Cream warmed, one pound of Currans well pickt, as much flower as will make it up, mingle them well together, make them into Cakes, bake them in an Oven; almost as hot as for Manchet, half an hour will bake them.

To make a Cake the way of the Royal Princess, the Lady Elizabeth, *daughter to King* Charles *the first.*

Take half a peck of Flower, half a pint of Rose-water, a pint of Ale-yeast, a pint of Cream, boil it, a pound and an half of Butter, six Eggs, (leave out the whites) four pound of Currans, one half pound of Sugar, one Nutmeg, and a little Salt, work it very well, and let it stand half an hour by the fire, and then work it again, and then make it up, and let it stand an hour and a half, in the Oven; let not your Oven be too hot.

To make Paste of Apricocks.

Take your Apricock, & pare them, and stone them, then boil them tender betwixt two dishes on a Chafing-dish of coals; then being cold, lay it forth on a white sheet of paper; then take as much sugar as it doth weigh, & boil it to a candy height, with as much Rose-water and fair water as will melt the sugar; then put the pulp into the Sugar, and so let it boil till it be as thick as for Marmalet, now and then stirring of it; then fashion it upon a Pye-plate like to half Apricocks, and the next day close the half Apricocks to the other, and when they are dry, they will be as cleer as Amber, and eat much better than Apricocks itself.

To make Paste of Pippins like leaves, and some like Plums, with their stones, and Stalks in them.

Take Pippins pared and coared, and cut in pieces, and boiled tender, so strain them, and take as much Sugar as the Pulp doth weigh, and boil it to a Candy height with as much Rose-water and fair water as will melt it, then put the pulp into the hot sugar, and let it boil until it be as thick as Marmalet; then fashion it on a Pye-plate, like Oaken leaves, and some like half Plums, the next day close the half Plums together; and if you please you may put the stones and stalks in them, and dry them in an Oven, and if you will have them look green, make the paste when Pippins are green; and if you would have them look red, put a little Conserves of Barberries in the Paste, and if you will keep any of it all the year, you must make it as thin as Tart stuff, and put it into Gallipots.

To make Paste of Elecampane roots, an excellent remedy for the Cough of the Lungs.

Take the youngest Elecampane roots, and boil them reasonably tender; then pith them and peel them; and so beat it in a Mortar, then take twice as much sugar as the Pulp doth weigh, and so boil it to a Candy height, with as much Rose-water as will melt it; then put the pulp into the Sugar with the pap of a roasted-apple, then let it boil till it be thick, then drop it on a Pye-plate, and so dry it in an Oven till it be dry.

To make Paste of flowers of the colour of Marble, tasting of natural flowers.

Take every sort of pleasing Flowers, as Violets, Cowslips, Gillyflowers, Roses or Marigolds, and beat them in a Mortar, each flower by it self with sugar, till the sugar become the colour of the flower, then put a little Gum Dragon steept in water into it, and beat it into a perfect paste; and when you have half a dozen colours, every flower will take of his nature, then rowl the paste therein, and lay one piece upon another, in mingling sort, so rowl your Paste in small rowls, as big and as long as your finger, then cut it off the bigness of a small Nut, overthwart, and so rowl them thin, that you may see a knife through them, so dry them before the fire till they be dry.

To make Paste of Rasberries or English Currans.

Take any of the Frails, and boil them tender on a Chafing-dish of coals betwixt two dishes and strain them, with the pap of a rosted Apple; then take as much sugar as the Pulp doth weigh, and boil to a Candy height with as much Rose-water as will melt it; then put the Pulp into the hot Sugar, and let it boil leisurely till you see it is as thick as Marmalet, then fashion it on a Pie-plate, and put it into the Oven with two billets of wood, that the place touch not the bottom, and so let them dry leasurely till they be dry.

To make Naples Bisket.

Take of the same stuff the Mackaroons are made of, and put to it an ounce of pine-apple-seeds in a quarter of a pound of stuff, for that is all the difference between the Mackaroons and the Naples Bisket.

To make Italian Biskets.

Take a quarter of a pound of searsed sugar, and beat it in an Alablaster mortar with the white of an Egg, and a little Gum Dragon steept in Rose-water, to bring it to a perfect paste, then mould it up with a little Anniseed and a grain of Musk; then make it up like Dutch-bread, and bake it on a Pie-plate in a warm Oven till they rise somewhat high and white, take them out, but handle them not till they be throughly dry and cold.

To make Prince Biskets

Take a pound of searsed sugar, and a pound of fine flower, eight Eggs with two of the reddest yolks taken out, and so beat together one whole hour, then take you Coffins, and indoice them over with Butter very thin, then put an ounce of Anniseeds finely dusted, and when you are ready to fill your Coffins, put in the Anniseeds and so bake it in an Oven as hot as for Manchet.

To make Marchpane to Ice and Gild, and garnish it according to Art.

Take Almonds, and blanch them out of seething water, and beat them till they come to a fine paste in a stone Mortar, then take fine searsed sugar, and so beat it altogether till it come to a prefect paste, putting in now and then a spoonful of Rose-water, to keep it from oyling; then cover your Marchpane with a sheet of paper as big as a Charger, then cut it round by that Charger, and set an edge about it as about a Tart, then bottom it with Wafers, then bake it in an Oven, or in a Baking-pan, and when it is hard and dry, take it out of the Oven, and ice it with Rose-water and Sugar, and the white of an Egg, being as thick as butter, and spread it over thin with two or three feathers; and then put it into the Oven again, and when you see it rise high and white, take it out again and garnish it with some pretty conceit, and stick some long Comfits upright in it, so gild it, then strow Biskets and Carrawayes on it. If your Marchpane be Oyly in beating, then put to it as much Rose-water as will make it almost as thin as to ice.

Lozenges

Take Blossoms of Flowers, and beat them in a bowl-dish, and put them in as much clarified Sugar as may come to the colour of the cover, then boile them with stirring, till it is come to Sugar again; then beat it fine, and searse it, and so work it up to paste with a little Gum Dragon, steep it in Rose-water, then print it with your mould, and being dry, keep it up.

To make Walnuts artificial.

Take searsed Sugar, and Cinnamon, of quantity a like, work it up with a little Gum Dragon, steep it in Rose-water, and print it in a mould made like a Walnut-shell, then take white Sugar Plates, print it in a mold made like a Walnut kernel, so when they are both dry, close them up together with a little Gum Dragon betwixt, and they will dry as they lie.

To make Collops like Bacon of Marchpane.

Take some of your Marchpane Paste, and work it in red Saunders till it be red; then rowl a broad sheet of white Paste, and a sheet of red Paste, three of the white, and four of the red, and so one upon another in mingled sorts, every red between, then cut it overthwart, till it look like Collops of Bacon, then dry it.

To make artificial Fruits.

Take a Mould made of Alablaster, three yolks, and tye two pieces together, and lay them in water an hour, and take as much sugar as will fill up your mold, and boil it in a *Manus Christi*, then pour it into your mould suddenly, and clap on the lid, round it about with your hand, and it will be whole and yellow, then colour it with what colour you please, half red, or half yellow, and you may yellow it with a little Saffron steept in water.

Touching Preserves and Pomanders.

To make an excellent perfume to burn between two Rose leaves.

Take an ounce of Juniper, an ounce of Storax, half a dozen drops of the water of Cloves, six grains of Musk, a little Gum Dragon steept in water, and beat all this to paste, then roll it in little pieces as big as you please, then put them betwixt two Rose-leaves, and so dry them in a dish in an Oven, and being so dried, they will will burn with a most pleasant smell.

To make Pomander.

Take an ounce of Benjamin, an ounce of Storax, and an ounce of Laudanum, heat a Mortar very hot, and beat all these Gums to a perfect paste; in beating of it, put in six grains of Musk, four grains of Civet; when you have beaten all this to a fine paste with you hands with Rose-water, rowl it round betwixt your hands, and make holes in the heads, and so string them while they be hot.

To make an Ipswich Water.

Take a pound of fine white Castle-soap shave it thin in a pint of Rose-water, and let it stand two or three days; then pour all the water from it, and put to it half a pint of freshwater; and so let it stand one whole day, then pour out that, and put half a pint more, and let it stand a night more then put to it half an ounce of powder called sweet Marjoram, a quarter of an ounce of the powder of Winter-Savory, two or three drops of the Oyl of Spike, and the Oyl of Cloves, three grains of Musk, and as much Ambergreese; work all these together in a fair Mortar, with the powder of an Almond Cake dryed, and beaten as small as fine flour, so rowl it round in your hands in Rose-water.

To make a sweet Smell.

Take the Maste of a sweet Apple-tree, being gathered betwixt the two Lady-dayes, and put to it a quarter of Damask Rose-water, & dry it in a dish in an Oven; wet in drying two or three times with Rose-water, then put to it an ounce of Benjamin, an ounce of Storax Calamintæ: these Gums being beaten to powder, with a few leaves of Roses, then you may put what cost of Smells you will bestow, as much Civet or Ambergreese, and beat it altogether in a Pomander or a Bracelet.

Touching Wine.

To make Hypocras.

Take four Gallons of Claret Wine, eight ounces of Cinnamon, three Oranges, of Ginger, Cloves, and Nutmegs a small quantity, Sugar six pound, three sprigs of Rosemary, bruise all the spices somewhat small, and so put them into the Wine, and keep them close stopped, and often shaked together a day or two, then let it run through a gelly bag twice or thrice with a quart of new Milk.

The Lady Thornburghs Syrup of Elders.

Take Elder-berries when they be red, bruise them in a stone Mortar, strain the juyce, and boil it to a consumption of almost half, scum it very clear, take it off the fire whilest it is hot, put in sugar to the thickness of a syrup; put it no more on the fire, when it is cold, put it into Glasses, not filling them to the top, for it will work like Beer.

This cleanseth the stomach and spleen, and taketh away all obstructions of the Liver, by taking the quantity of a spoonful in a morning, and fasting a short time after it.

To make gelly of Raspis the best way.

Take the Raspis, and set them over the fire in a Posnet, and gather out the thin juyce, the bottom of the skillet being cooled with fair water, and strain it with a fine strainer, and when you have as much as you will, then weigh it with Sugar, and boil them till they come to a Gelly, which you may perceive by drawing your finger on the back of the spoon.

To dry Fox Skins.

Take your shee Fox Skins, nail them upon a board as strait as you can, then brush them as clean as you can, then take Aqua Fortis, and put into it a six pence, and still put in more as long as it will dissolve it, then wash your skin over with this water, and set it to dry in the sun; and when it is dry, wash it over with the spirits of wine; this must be done in hottest time of Summer.

Choice Secrets made known.

To make true Magistery of Pearl.

Dissolve two or three ounces of fine seed Pearl in distilled Vinegar, & when it is perfectly dissolved, and all taken up, pour the Vinegar into a clean glass Bason; then drop some few drops of Oyl of Tartar upon it, & it will cast down the Pearl into fine Powder, then pour the Vinegar clean off softly, then put to the Pearl clear Conduit or Spring water; pour that off, and do so often untill the taste of the Vinegar and Tartar be clean gone, then dry the powder of Pearl upon warm embers, and keep it for your use.

How to make Hair grow.

Take half a pound of Aqua Mellis in the Spring time of the year, warm a little of it every Morning when you rise in a Sawcer, and tie a little spunge to a fine box comb, and dip it in the water, and therewith moisten the roots of the Hair in combing it, and it will grow long, thick, and curled in a very short time.

To write Letters of Secret, that they cannot be read without the directions following.

Take fine Allum, beat it small, and put a reasonable quantity of it into water, then write with the said water.

The work cannot be read, but by steeping your paper in fair running water.

You may likewise write with Vinegar, or the juyce of Lemon or Onion; if you would read the same, you must hold it before the fire.

How to keep Wine from Sowring.

Tye a piece of very salt Bacon on the inside of your barrel, so as it touch not the Wine, which will preserve Wine from sowring.

To take out Spots of Grease or Oyl.

Take bones of sheeps feet, burn them almost to ashes, then bruise them to powder, and put of it on the spot, and lay it in the sun when it shineth hottest, when the powder becomes black, lay on fresh in the place till it fetch out the spots, which will be done in a very short time.

To make hair grow black, though any colour.

Take a little Aqua Fortis, put therein a groat or sixpence, as to the quantity of the aforesaid water, then set both to dissolve before the fire, then dip a small spunge in the said water, and wet your beard or hair therewith; but touch not the skin.

King Edwards *perfume.*

Take twelve spoonfuls of right red Rose-water, the weight of six pence in fine powder of Sugar, and boil it on hot Embers and Coles softly, and the house will smell as though it were full of Roses; but you must burn the sweet Cypress wood before, to take away the gross air.

Queen Elizabeths *Perfume.*

Take eight spoonfuls of Compound water, the weight of two pence in fine powder of Sugar, and boil it on hot Embers and Coals, softly, and half an ounce of sweet Marjoram dried in the Sun, the weight of two pence of the powder of Benjamin. This Perfume is very sweet, and good for the time.

Mr. Ferene *of the* New Exchange, *Perfumer to the Queen, his rare Dentifrice, so much approved of at Court.*

First take eight ounces of Ireos roots, also four ounces of Pomistone, and eight ounces of Cutle-bone, also eight ounces of Corral, and a pound of Brick if you desire to make them red; but he did oftener make them white, and then instead of the Brick did take a pound of fine Alabaster; all this being throughly beaten, and sifted through a fine searse, the powder is then ready prepared to make up in a paste, which must be done as follows.

To make the said Powder into Paste.

Take a little Gum Dragant, and lay it in steep twelve hours, in Orange flower water, or Damask Rose-water, and when it is dissolved, take the sweet Gum, and grind it on a Marble stone with the aforesaid powder, and mixing some crums of white bread, it will come into a Paste, the which you may make Dentifrices, of what shape or fashion you please, but rolls is the most commodious for your use.

The Receipt of the Lady Kents *powder, presented by her Ladyship to the Queen.*

Take white Amber, Crabs eyes, red Corral, Harts-horn and Pearl, all prepared several, of each a like proportion, tear and mingle them, then take Harts-horn gelly, that hath some Saffron put into a bag, dissolve into it while the gelly is warm, then let the gelly cool, and therewith make a paste of the powders, which being made up into little balls, you must dry gently by the fire side. Pearl is prepared by dissolving it with the juyce of Lemons, Amber prepared by beating it to powder; so also Crabs-eyes and Coral, Harts-horn prepared by burning it in the fire, and taking the shires of it especially, the pith wholly rejected.

A Cordial Water of Sir Walter Raleigh.

Take a gallon of Strawberries, and put them into a pint of *Aqua vitæ*, let them stand for four or five days, strain them gently out, and sweeten the water as you please with fine Sugar; or else with perfume.

The Lady Malets *Cordial Water.*

Take a pound of fine Sugar beaten and put to it a quart of running water, pour it three or four times through a bag; then put a pint of Damask Rose-water, which you must always pour still through the bag, then four penniworth of Angelica water, four pence in Clovewater, four pence of Rosa Solis, one pint of Cinnamon-water, or three pints and a half *Aqua vitæ*, as you find it in taste; put all these together three or four times through the bag or strainer, and then take half an ounce of good Muskallis and cut them grosly, & put them into a glass, and fill them with the water, &c.

A Sovereign Water of Dr. Stephens, *which he long times used, wherewith he did many Cures; he kept secretly till a little before his death, and then he gave it to the Lord Arch-bishop of* Canterbury *in writing, being as followeth,* viz.

Take a Gallon of good Gascoine Wine, and take Ginger, Gallingale, Cinamon, Nutmegs, Cloves, Grains, Anniseeds, Fennilseed, of every of them a dram, then take Caraway-seed, of red Mints, Roses, Thime, Pellitory of the Wall, Rosemary, wild Thime, Camomil, the leaves if you cannot get the flowers, of small Lavander, of each a handful, then bray the Spices small, and bray the Herbs, and put all into the Wine, and let it stand for twelve hours, stirring divers times, then still it in a Limbeck, and keep the first water, for it is best, then put the second water by it self, for it is good, but not of such vertues, &c.

The Vertues of this water.

It comforts the Spirits Vital, and helps all inward Diseases that come of cold, it is good against the shaking of the Palsie; it cures the contraction of the Sinews, helps the conception of Women if they be Barren, it kills the Worms in the Belly and Stomach; it cures the cold Dropsie, and helps the Stone in the Bladder, and in the Reins of the back; it helps shortly the stinking breath, and whosoever useth this Water morning and evening, (and not too often) it preserveth him in good liking, and will make him seem young very long, and Comforteth nature marvellously; with this water did Dr. *Stephens* preserve his life, till extream age would not let him go or stand and he continued five years, when all the Physicians judged he would not live a year longer, nor did he use any other Medicine but this, &c.

A Plague Water to be taken one spoonful every four hours with one sweat every time.

Take Scabious; Betony, Pimpernel, and Turmentine-roots, of each a pound, steep these all night in three gallons of strong Beer, and distil them all in a Limbeck, and when you use it, take a spoonful thereof every four hours, and sweat well after it, draw two quarts of water, if your Beer be strong, and mingle them both together.

Poppy water.

Take four pound of the flower of Poppies well pickt and sifted, steep them all night in three Gallons of Ale that is strong, and still it in a Limbeck; you may draw two quarts, the one will be strong and the other will be small, &c.

A Water for a Consumption, or for a Brain that is weak.

Take Cream (or new milk) and Claret-wine, of each three pints of Violet-flowers, Bugloss and Borage-flowers, of each a spoonful, Comfrey, Knot-grass, and Plantane of these half a handful, three or four Pome-waters sliced, a stick of Liquorish, some Pompion seeds and strings; put to this a Cock that hath been chased and beaten before he was killed, dress it as to boil, and parboil it until there be no blood in it; then put them in a pot, and set them over your Limbeck, and the soft fire; draw out a pottle of water, then put your water in a Pipkin over a Charcoal fire, and boil it a while, dissolve therein six ounces of white Sugar-candy, & two penny weight of Saffron: when it is cold strain it into a glass, & let the Patient drink three or four spoonfuls three or four times a day blood-warm; your Cock must be cut into small pieces, & the bones broken, and in case the flowers and herbs are hard to come by, a spoonful of their stilled waters are to be used.

Another of the same.

Take a pottle of good Milk, one pint of Muscadine, half a pint of red Rose-water, a penny manchet sliced thin, two handfuls of Raisins of the sun stoned, a quarter of a pound of fine sugar, sixteen Eggs beaten; mix all these together, then distill them in a common still with a soft fire, then let the Patient drink three or four spoonfuls at a time blood warm, being sweetned with *Manus Christi* made with Corral and Pearl; when your things are all in the still, strew four ounces of Cinamon beaten; this water is good to put into broath, &c.

A good Stomach Water.

Take a quart of *Aqua Composita*, or *Aqua vitæ*, (the smaller) and put into it one handful of Cowslip flowers, a good handful of Rosemary flowers, sweet Marjoram, a little Pellitory of the Wall, a little Betony and Balm, of each a little handful, Cinnamon half an ounce, Nutmegs a dram, Anniseeds, Coriander seeds, Caroway seeds, Gromel seeds, Juniper berries, of each a dram, bruise the spice and seed, and put them into *Aqua Composita*, or *Aqua vitæ*, with your Herbs together, and put into them a pound of very fine sugar, stir them well together, and put them into a glass and let it stand in the sun nine days, and stir it every day; two or three Dates, and a little race of Ginger sliced into it will make it the better, especially against wind, &c.

A Bag of purging Ale.

Take of Agrimony, Speedwell, Liverwort, Scurvy-grass, Water-cresses, of each a handful, of Monks Rhubarb, and red Madder, of each half a pound, of Horseradishes three ounces, Liquorish two ounces, Sassafrage four ounces, Sena seven ounces, sweet Fennil-seeds two drams, Nutmegs four; pick and wash your Herbs and Roots, and bruise them in a Mortar, and put them in a bag made of a Bolter, & so hang them in three gallons of middle Ale, and let it work in the Ale, and after three days you may drink it as you see occasion, &c.

The Ale of Health and Strength, by Viscount St. Albans.

Take Sassafras wood half an ounce, Sarsaparilla three ounces, white Saunders one ounce, Chamapition an ounce, China-root half an ounce, Mace a quarter of an ounce, cut the wood as thin as may be with a knife into small peices, and bruise them in a Mortar; put to them these sorts of Herbs, (viz.) Cowslip flowers, Roman-wormwood, of each a handful, of Sage, Rosemary, Betony, Mug-wort, Balm and Sweet-marjoram, of each half a handful, of Hops; boil all these in six gallons of Ale till it come to four; then put the wood and hearbs into six gallons of Ale of the second wort, and boil it till it come to four, let it run from the dregs, and put your Ale together, and tun it as you do other purging Ale, &c.

A Water excellent good against the Plague.

Take three pints of Malmsey, or Muscadine, of Sage and Rue, of each one handful, boil them together gently to one pint, then strain it and set it on the fire again, and put to it one penniworth of Long Pepper, Ginger four drams, Nutmegs two drams, all beaten together, then let it boil a little, take it off the fire, and while it is very hot, dissolve therein six penniworth of Mithridate, and three penniworth of Venice Treacle, and when it is almost cold put to it a pint of strong Angelica water, or so much *Aqua vitæ*, and so keep it in a glass close stopped.

A Cordial Cherry-water.

Take a pottle of *Aqua vitæ*, two ounces of ripe Cherries stoned, Sugar one pound, twenty four Cloves, one stick of Cinamon, three spoonfuls of aniseeds bruised, let these stand in the *Aqua vitæ* fifteen days, and when the water hath fully drawn out the tincture, pour it off into another glass for your use, which keep close stopped, the Spice and the Cherries you may keep, for they are very good for winde in the Stomach.

The Lord Spencers Cherry-water.

Take a pottle of new Sack, four pound of through ripe Cherries stoned, put them into an earthen pot, to which put an ounce of Cinnamon, Saffron unbruised one dram, tops of Balm, Rosemary or their flowers, of each one handful, let them stand close covered twenty four hours, now and then stirring them; then put them into a cold Still, to which put of beaten Amber two drams, Corianderseed one ounce, Alkerms one dram, and distill it leisurely, and when it is fully distilled, put to it twenty grains of Musk. This is an excellent Cordial, good for Faintings and Swoundings, for the Crudities of the Stomach, Winde and Swelling of the Bowels, and divers other evil Symptomes in the Body of Men and Women.

The Herbs to be distilled for Usquebath.

Take Agrimony, Fumitory, Betony, Bugloss, Wormwood, Hartstongue, Carduus Benedictus, Rosemary, Angelica, Tormentil, of each of these for every gallon of Ale one handful, Anniseed, and Liquorish well bruised half a pound, still these together, and when it is stilled, you must infuse Cinamon, Nutmeg, Mace, Liquorish, Dates, and Raisins of the Sun, and sugar what quantity you please. The infusion must be till the colour please you.

Dr. Kings *way to make Mead.*

Take five quarts and a pint of water, and warm it, then put one quart of Honey to every gallon of Liquor, one Lemon, and a quarter of an ounce of Nutmegs; it must boil till the scum rise black, that you will have it quickly ready to drink, squeeze into it a Lemon when you tun it. It must be cold before you tun it up.

To make Syrup of Rasberries.

Take nine quarts of Rasberries, clean pickt, and gathered in a dry day, and put to them four quarts of good Sack, into an earthen pot, then paste it up very close, and set it in a Cellar for ten days, then distill it in a Glass or Rosestill, then take more Sack and put in Rasberries to it, then when it hath taken out all the colour of the Raspis, strain it out and put in some fine Sugar to your taste, and set it on the fire, keeping it continually stirring till the scum doth rise; then take it off the fire, let it not boil, skim it very clean, and when it is cold put it to your distilled Raspis; colour it no more than to make it a pale Claret Wine. This put into bottles or Glasses stopt very close.

To make Lemon Water.

Take twelve of the fairest Lemons, slice them, and put them into two pints of White wine, and put to them Cinamon two drams, Gallingale two drams, of Rose-leaves, Borage and Bugloss flowers, of each one handful, of yellow Saunders one dram; steep all these together twelve hours; then distill them gently in a Glass still untill you have distilled one pint and an half of the Water, and then adde to it three ounces of Sugar; one grain of Ambergreese, and you will have a most pleasing cleansing Cordial water for many uses.

To make Gilly-flower Wine.

Take two ounces of dryed Gilly-flowers, and put them into a pottle of Sack, and beat three ounces of Sugar-candy, or fine Sugar and grind some Ambergreese, and put it in the bottle and shake it oft, then run it through a gelly bag, and give it for a great Cordial after a weeks standing or more. You may make Lavander as you do this.

The Lady Spotswood Stomach Water.

Take white Wine one pottle, Rosemary and Cowslip flowers, of each one handful, as much Betony leaves, Cinamon and Cloves grosly beaten, of both one ounce; steep all these three dayes, stirring it often; then put to it Mithridate four ounces, and stir it together, and distil it in an ordinary still.

Water of Time for the Passion of the Heart.

Take a quart of white Wine, and a pint of Sack, steep in it as much broad Thime as it will wet, put to it of Galingale and Calamus Aromaticus, of each one ounce, Cloves, Mace, Ginger, and grains of Paradise two drams, steep these all night, the next morning distil it in an ordinary still, drink it warm with Sugar.

A Receipt to make damnable Hum.

Take Species de Gemmis, Aromaticum Rosatum, Diarrhodon Abbatis, Lætificans Galeni, of each four drams, Loaf-sugar beaten to powder half a pound, small *Aqua Vitæ* three Pints, strong Angelica water one pint; mix all these together, and when you have drunk it to the Dregs, you may fill it up again with the same quantity of water. The same powders will serve twice, and after twice using it, it must be made new again.

An admirable Water for sore Eyes.

Take *Lapis Tutiæ*; Aloes Hepatica, fine hard sugar, of each three drams, beat them very small, and put them into a Glass of three pints, to which put red Rose-water and white Wine, of each one pint; set the Glass in the Sun, in the Month of *July*, for the whole Month, shaking it twice in a day for all that while; then use it as followeth, put one drop thereof into the Eye in the evening, when the party is in bed, and one drop in the morning an hour before the Patient riseth: Continue the use of it till the Eyes be well. The older the Water, the better it is. Most approved.

A Snail Water for weak Children, and old People.

Take a pottle of Snails, and wash them well in two or three waters, and then in small Beer, bruise them shells and all, then put them into a gallon of red Cows Milk, red Rose leaves dried, the whites cut off, Rosemary, sweet Marjoram, of each one handful, and so distil them in a cold still, and let it drop upon powder of white

Sugar candy in the receiver; drink of it first and last, and at four a clock in the afternoon, a wine-glass full at a time.

Clary Water for the Back, Stomach, &c.

Take three gallons of midling Beer, put in a great brass Pot of four gallons, and put to it ten handfuls of Clary gathered in a dry day, Raisins of the Sun stoned three pounds, Anniseeds, and Liquorish, of each four ounces, the whites and shells of twenty four eggs, or half so many, if there be not so much need, beat the shells small, and mix them with the whites; put to the bottoms of three white loaves, put into the Receiver one pound of white sugar-candy, or so much fine loaf sugar beaten small, and distill it through a Limbeck, keep it close, and be seldom without it; for it reviveth very much the stomach and heart, strengtheneth the back, procureth appetite and digestion, driveth away Melancholly, sadness and heaviness of the heart, &c.

Dr. Montfords *Cordial Water.*

Take Angelica leaves twelve handfuls, six leaves of Carduus Benedictus, Balm & Sage, of each five handfuls, the seeds of Angelica and sweet Fennil, of each five ounces bruised, scraped and bruised Liquorish twelve ounces, Aromaticum Rosatum, Diamoscus dulcis, of each six drams; the Herbs being cut small, the seeds and Liquorish bruised, infuse them into two gallons of Canary Sack for twenty four hours, then distill it with a gentle fire, and draw off onely five pints of the spirits, which mix with one pound of the best Sugar dissolved into a Syrup in half a pint of pure red Rose-water.

Aqua Mirabilis, Sir Kenelm Digby's *way.*

Take Cubebs, Gallingale, Cardamus, Melliot flowers, Cloves, Mace, Ginger, Cinamon, of each one dram bruised small, juyce of Celandine one pint, juyce of Spearmint half a pint, juyce of Balm half a pint, sugar one pound, flower of Cowslips, Rosemary, Borage, Bugloss, Marigolds, of each two drams, the best Sack three pints, strong Angelica water one pint, red Rose-water half a pint, bruise the Spices and flowers, & steep them in the Sack & Juyces one night, the next morning distill it in an ordinary Limbeck or glass still, and first lay Hearts-tongue leaves in the bottom of the Still.

The Vertues of the precedent Water.

This Water preserveth the Lungs without grievances, and helpeth them; being wounded, it suffereth the blood not to putrifie, but multiplieth the same; this water suffereth not the heart to burn, nor melancholly, nor the Spleen to be lifted up above nature; it expelleth the Rheum, preserveth the Stomach, conserveth Youth, and procureth a good colour, it preserveth Memory, it destroyeth the palsie; if this be given to one a dying, a spoonful of it reviveth him; in the summer use one spoonful a week fasting, in the winter two spoonfuls.

A Water for fainting of the Heart.

Take Bugloss and red Rose-water of each one pint, Milk half a pint, Anniseeds and Cinamon grosly bruised, of each half an ounce, Maiden-hair two handfuls, Harts-tongue one handful, both shred, mix all together, and distill it in an ordinary still, drink of it morning and evening With a little sugar.

A Surfeit Water.

Take half a bushel of red Corn Poppy, put it into a large dish, cover it with brown Paper, and lay another dish upon it, set it in an Oven after brown bread is baked divers times till it be dry, which put into a pottle of good *Aqua vitæ*, to which put Raisins of the sun stoned half a pound, six figs sliced, three Nutmegs sliced, two flakes of Mace bruised, two races of Ginger sliced, one stick of Cinnamon bruised, Liquorish sliced one ounce, Aniseed, Fennil-seed, and Cardamums bruised, of each one dram; put all these into a broad glass body, and lay first some Poppy in the bottom, then some of the other ingredients, then Poppy again, and so untill the Glass be full; then put in the *Aqua vitæ*, and let it infuse till it be strong of the spices, and very red with the Poppy, close covered, of which take two or three spoonfuls upon a surfeit, and when all the liquor is spent, put more *Aqua vitæ* to it, and it will have the same effect the second time, but no more after.

Dr. Butlers *Cordial Water against Melancholly, &c. most approved.*

Take the flowers of Cowslips, Marigolds, Pinks, Clove-gilly-flowers, single stock gilly-flowers, of each four handfuls, the flowers of Rosemary, and Damask Roses, of each three handfuls, Borage and Bugloss flowers, and Balm leaves, of each two handfuls; put them in a quart of Canary Wine into a great Bottle or Jug close stopped, with a Cork, sometimes stirring the flowers and wine together, adding to them Anniseeds bruised one dram, two Nutmegs sliced, *English* Saffron two pennyworth; after some time of infusion, distill them in a cold Still with a hot fire, hanging at the Nose of the Still Ambergreece and Musk, of each one grain; then to the distilled water put White Sugar-candy finely beaten six ounces, and put the glass wherein they are into hot water for one hour. Take of this water at one time three spoonfuls thrice a week, or when you are ill, it cureth all melancholly fumes, and infinitely comforts the spirits.

The admirable and most famous Snail Water.

Take a peck of garden shell snails, wash them well in small beer, and put them in a hot Oven till they have done making a noise, then take them out, and wipe them well from the green froth that is upon them, and bruise them shells and all in a stone Mortar, then take a quart of earth worms, scower them with salt, slit them & wash them well with water from their filth, and in a stone Mortar beat them to pieces, then lay in the bottom of your distilled pot Angelica two handfuls, and two handfuls of Celandine upon them, to which put two quarts of Rosemary flowers, Bears foot, Agrimony, red Dock Roots, Bark of Barberries, Betony, Wood sorrel, of each two handfuls, Rue one handful; then lay the Snails and worms on the top of the Herbs and Flowers, then pour on three Gallons of the strongest Ale, and let it stand all night, in the morning put in three ounces of Cloves beaten, six penniworth of beaten Saffron and on the top of them six ounces of shaved Harts-horn, then set on the Limbeck, and close it with paste, and so receive the water by pints, which will be nine in all, the first is the strongest, whereof take in the morning two spoonfuls in four spoonfuls of small Beer, and the like in the afternoon; you must keep a good Diet and use moderate exercise to warm the blood.

This Water is good against all Obstructions whatsoever. It cureth a Consumption and Dropsie, the stopping of the Stomach and Liver. It may be distilled with milk for weak people and children, with Harts-tongue and Elecampance.

A singular Mint water.

Take a still full of Mints, put Balm, and Penniroyal, of each one good handful, steep them in Sack, or Lees of Sack twenty four hours, stop it close, and stir it now and then: Distill it in an ordinary Still with a very quick fire, and keep the Still with wet cloaths, put into the receiver as much sugar as will sweeten it, and so double distill it.

Distillings.

A most Excellent **Aqua Coelestis** *taught by Mr.* **Philips Apothecary.**

Take of Cinamon one dram, Ginger half a dram, the three sorts of Saunders, of each of them three quarters of an ounce, Mace and cubebs, of each of them one dram, Cardamom the bigger and lesser, of each three drams, Setwall-roots half an ounce, Anniseed, Fennilseed Basil-seed, of each two drams, Angelica roots, Gilly-flowers, Thyme, Calamint, Liquorish, Calamus, Masterwort, Pennyroyal, Mint, Mother of Thyme, Marjoram, of each two drams, red Roseseed, the flowers of Sage and Betony, of each a dram and a half, Cloves, Galingal, Nutmegs, of each two drams, the flowers of Stechados, Rosemary, Borage and Bugloss flowers, of each a dram and half, Citron Rindes three drams; bruise them all, and put in these Cordial Powders, Diamber Aromaticum, Diamascum, Diachoden, the Spices made with Pearl, of each three drams; infuse all these in twelve pints of *Aqua Vitæ*; in a glass, close stopped for fifteen dayes, often shaking it, then let it be put into a Limbeck close stopped, and let it be distilled gently; when you have done, hang in a cloth, two drams of Musk, half a dram of Ambergreese, and ten or twelve grains of gold, and so receive it to your use.

Hypocras taught by Dr. Twine for Wind in the Stomach.

Take Pepper, Grains, Ginger, of each half an ounce, Cinnamon, Cloves, Nutmegs, Mace, of each one ounce grosly beaten, Rosemary, Agrimony, both shred of each a few crops, red Rose leaves a pretty quantity, as an indifferent gripe, a pound of Sugar beaten; lay these to steep in a gallon of good Rhenish or white-Wine in a close vessel, stirring it two or three times a day the space of three or four dayes together, then strain it through an Hypocras strainer, and drink a draught of it before meat half an hour, and sometimes after to help digestion.

Marigold flowers distilled, good for the pain of the Head.

Take Marigold flowers, and distill them, then take a fine cloth and wet in the aforesaid distilled water, and so lay it to the forehead of the Patient, and being so applied, let him sleep if he can; this with Gods help will cease the pain.

A Water good for Sun burning.

Take Water drawn off the Vine dropping, the flowers of white Thorn, Bean-flowers, Water Lilly-flowers, Garden Lilly-flowers, Elder-flowers, and Tansie-flowers, Althea-flowers, the whites of Eggs, French Barley.

The Lady Giffords cordial Water.

Take four quarts of *Aqua vitæ*, Borrage and Poppy-water, of each a pint, two pounds of Sugar-candy, one pound of figs sliced, one pound of Raisins of the Sun stoned, two handfuls of red Roses clipped and dried, one handful of red Mint, half a handful of Rosemary, as much of Hysop, a few Cloves; put all these in a great double Glass close stopped, and set it in the sun three months, and so use it.

A water for one pensive and very sick, to comfort the Heart very excellent.

Take a good spoonful of *Manus Christi*, beaten very small into powder, then take a quarter of a pound of very fine sugar, and beat it small, and six spoonfuls of Cinamon water, and put to it, and ten spoonfuls of red Rose-water; mingle all these together, and put them in a dish, and set them over a soft fire five or six walms, and so let it be put into a glass, and let the party drink thereof a spoonful or two, as he shall see cause.

To perfume Water.

Take Malmsey or any kind of sweet water; then take Lavender, Spike, sweet Marjoram, Balm, Orange peels, Thyme, Basil, Cloves, Bay leaves, Woodbine flowers, red and white Roses, and still them all together.

FINIS.

www.ingramcontent.com/pod-product-compliance
Lightning Source LLC
Chambersburg PA
CBHW030451220526
45464CB00006B/2494